Acknowledgment

I must thank my grandmother, Mrs. Julia Gray, for all her un-ending encouragement, and the Sleznicks, Carol and Irwin, for the motivation and support. Secondly, many people read earlier transcripts and made helpful suggestions. For all their contributions I would like to thank Jim Edwards and Paul DeBruyn for the early comments and Craig Flatten, Dick McNeely, Paul Meehan-Martin, Mary Mullen, Tom Stralser, Richard Easterly & Debra Salstrom for their contributions to this text. Thanks to Laurel Moony for her help with the layout. Thanks to Grant Beck and Dick Rice for all their efforts in making this possible. A special thanks to Bob Lewis and Graphiques House for putting it all together.

THE BARN OWL AND THE PELLET

By Bret Gaussoin
and Janice Lapsansky

Illustrated by
Andrew S. Bolland

This book is dedicated to my wife Kim, for without her this bird man would still be on page 1.

ISBN #0 9620281 0-X
Pellets, Inc. / P.O. Box 5484 / Bellingham/ WA / 98227-5484

Printed in the U.S.A.

Preface

Undoubtedly, the best way to learn about the living world is to invest considerable time exploring the beaches, forests, and fields which surround us. Many of us, however, have neither the time nor the opportunity to study wildlife in its natural state. Thus, we must learn about ourselves and other living organisms by other means. While books on biology and ecology provide such data, we may also look for clues to the ways of nature in what might seem to be the most unusual of places.

The barn owl, for example, can teach us many things about his habits even without the benefit of his physical appearance. The pellet is the final product of the interaction between this predator and its prey. It can serve as a window into life as it demonstrates the final link in the **food chain** that supports a barn owl's very existence. In short, each barn owl becomes an important participant in providing structure to the interweaving strands of the web of life.

Information obtained from this manual and the study of barn owl pellets can provide an interesting introduction to **ecological** structure and a consciousness for life science. It is intended to help explain the relationship between animal structure and function, as well as our role as a member of the **biosphere.** We hope this exercise will prove as interesting to you as it is to us; and hope it will stimulate further study and a greater appreciation for the wildlife in this world we share.

Table of Contents

Page

A denzien of the night

Barn Owl Folklore

The barn owl, like other predators, has suffered under a shroud of myth and superstition. Owls are well represented in the literature of the distant past, always sinister characters (denziens of the night) and often as messengers of death or misfortune. In the Biblical account of the fall of Babylon, the prophet Isaiah foretells of a city in which...

"their houses shall be full of doleful creatures; and **owls shall dwell there,** and satyrs shall dance there. . ."

On the other hand the owl has long commanded an image of wisdom. Alfred Lord Tennyson's well known poem, The **White Owl,** illustrates the somber yet shrewd character often attributed to owls:

"Alone and warming his five wits,
The white owl in the belfry sits."

In Britain it was a common belief that the barn owl carried a death summons to anyone whose house it perched upon and uttered strange cries. This myth was encouraged by poets and magicians of the time, and even Shakespeare used the owl in several of his plays as a symbol of impending death. For example, upon hearing the piercing cry of an owl, Lady Macbeth exclaims:

"Hark! Peace!
It was the owl that shrieked, the fatal bellman,
Which gave the stern'st good night."

But in John Gary's fabel, **The Shepherd and the Philosopher,** the author alludes to **both** the sinister and the wise:

"Can grave and formal pass for wise,
when men the solemn owl despise?"

Barn Owl Folklore

Man's earliest ecological inclinations cast all predators in the role of villain. They became known as selfish, merciless killers of less keen and agile creatures. But nature's methods, while harsh, were poorly understood and often feared. Thus, prior to the beginning of the 19th century, only those plants and animals proven beneficial or useful to man were among those worth protecting. Quite often these natural (or unnatural) resources were considered the private property of landowners and were left unmanaged.

Eventually the movement to protect hunting and fishing game led to a more conscientious and comprehensive approach to conservation. Science began to understand the valuable role each organism plays in the ecosystem. Indeed we are still learning that the preservation of threatened and endangered life and even our entire environment is accomplished only through a commitment in action and voice. And we must always remember . . . we too are part of the web of life.

The Pellet

What is an owl pellet? This blackish clod of fur or feathers and bone is the undigestible remains of small animals eaten by a wild barn owl, **Tyto alba** (Fig. 1). Because it usually swallows its **prey** whole and is able to digest only fleshy parts, the barn own must regurgitate the remaining solid material in the form of a compact pellet or **casting** (Fig.2). Though birds of many **species** regurgitate pellets, barn owl pellets are especially suited for study because they are large enough to be examined without the aid of a microscope; and they generally contain the entire skeletons of animals that are eaten. And because they accumulate in predictable locations, they are more readily available.

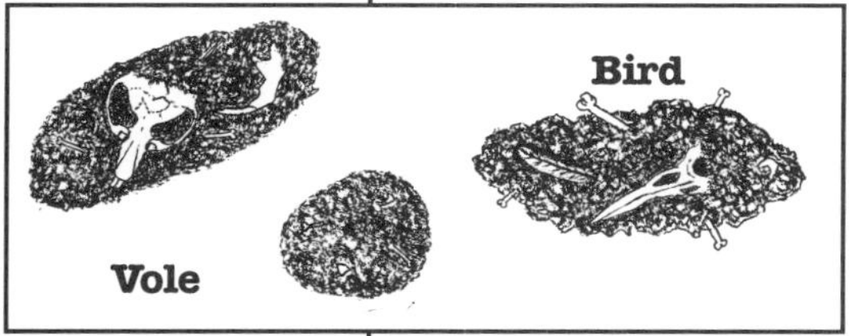

Figure 1.
Barn Owl Pellets

Figure 2.
Pellet formation. Prey is swallowed whole and proceeds to the stomach; hard parts are separated out and regurgitated.

The barn owl requires six to twelve hours after its last meal to form and pass a pellet. During this time it retires to a quiet perch to **preen** and doze. After the pellet has been cast, it (the casting) plays a new role in nature by becoming home to a variety of insects. The principal tenant of an aging owl pellet is the wool-eating clothes moth and its **larvae** (Fig.3), but carpet beetles may also feed on the fur and feathers within. Because these tiny animals live in and around owl pellets, so do the **predators** that eat them, primarily spiders. (Note: You have been provided with pellets that no longer contain any living insects.)

The Pellet

Dissection of a barn owl pellet will reveal the exact diet of a particular bird, and with careful study, much more. For example, the shape of the pellet can tell us if it landed on a floor or some other man-made structure or was dropped in a more natural environment, thus showing us the nearly exact habitat of the owl. The kind of plant material clinging to the outside of the pellet might also indicate that it was found among hay or straw in a barn loft, or amongst the grain of a silo. If it has been weathered by the wind and rain, or had been sheltered from the elements also shows the natural habitat of the particular barn owl. Bird feathers and droppings can be very distinctive, especially those of the pigeon, the barn owl's most common roommate in manmade structures.

Moth Larvae

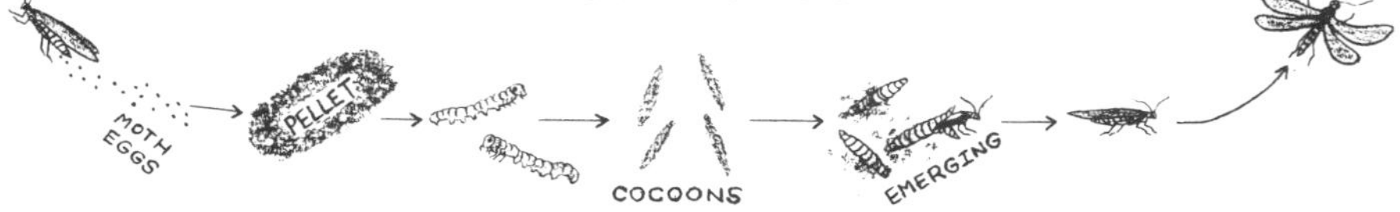

Figure 3.
This cycle of the wool-eating moth takes place entirely inside the owl pellet except for the free flying adult.

The bones hidden inside the pellet usually represent the entire skeleton of every animal that the owl has captured and eaten in a night of foraging. There are almost always remains of two or more animals present in each pellet. It is possible that only portions of animals were eaten, which will make identification more difficult. As you dissect the pellet, sort the similarly-sized bones and you may be able to partially reconstruct the skeletons of different prey animals. Skulls and jawbones, of course, are the easiest to identify. (Refer to the Picture Key)

Barn Owl Ecology

Ecology is the study of living organisms and their relationships with both their **biotic** and **abiotic** environments. This association has influence in both directions: the environment strongly influences the life of each organism, and the presence of every plant and animal continually alters the environment. Natural processes direct the flow of energy through the orderly arrangement of organisms in the **food web,** and support the integrity of the **ecosystem** (Fig. 4).

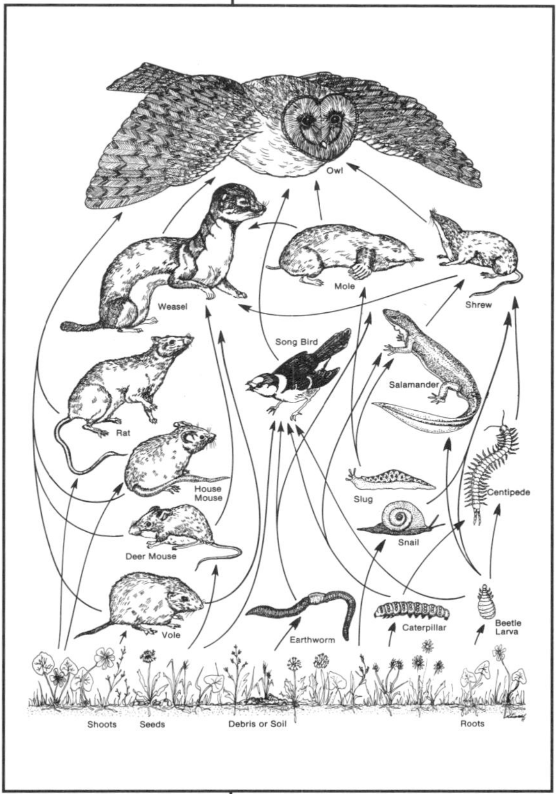

Figure 4.
An example of the food web.

The life of a barn owl may well be a perfect model of the simplest food chain: plant-herbivore-carnivore (Fig. 5). However, the transfer of energy through even the simplest food chain is never completely efficient. The grasses on which mice feed convert only a small percentage of the **solar energy** that strikes their leaves into plant chemical energy. And, in digesting plant material, only a small fraction of the stored **chemical energy** is available for use by the mice. Similarly, as energy for growth and locomotion is extracted by the owl from the chemical energy of mouse flesh, much energy is lost to the environment in the form of **metabolic** waste products and heat. Further, because matter can neither be created nor destroyed, the barn owl passes on both its stored energy and chemicals in death to the organisms called **decomposers.** Decomposers break down the remains of the owl, thereby making raw materials available to other organisms, thus maintaining the cycle.

Barn Owl Ecology

In the Pacific Northwest, the barn owl preys primarily on the common field mouse; it takes hundreds of mice to supply a single barn owl with its energy needs for one year. This ecological rule of incomplete energy transfer may be illustrated by constructing a **trophic pyramid,** where each level represents the total **biomass** comprising each link in the food chain needed to support the biomass of the next highest level. Because predators usually represent the final link in any food chain, they occupy the uppermost position in the trophic pyramid, an important but vulnerable status.

Figure 5.
Two simple food chains.

Since the barn owl occupies the highest level of its own trophic pyramid, it may be used as a **biological indicator.** A study of the **Tyto alba's** past and present population may well disclose the apparent health of the ecosystem. Changes in the availability of a primary food source due to drought, pollution, or other environmental stresses are reflected in the population size of the predator (Fig. 6). Since we humans are at the top of our trophic pyramid, we depend, for example, on the continued health of rangeland (the near bottom of our trophic pyramid) in order to produce enough plants to support grazing beef cattle.

Population Fluctuations

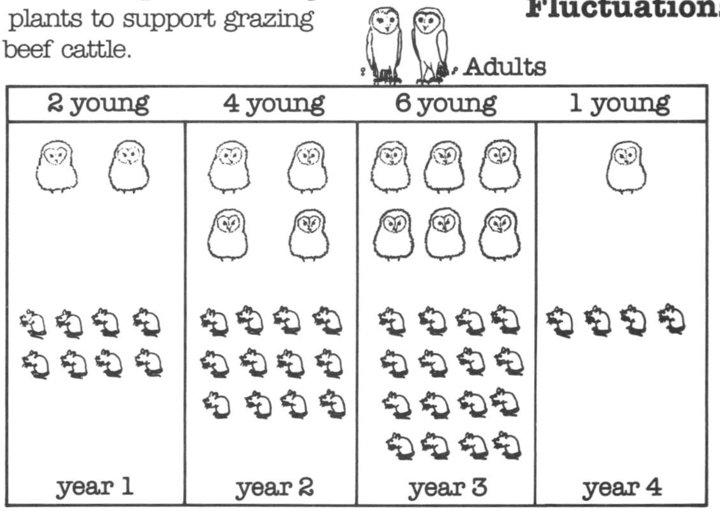

Figure 6.
Barn owl population fluctuations may coincide with those of their prey or other environmental factors.

Barn Owl Description & Adaptations

Habitat

The barn owl is considered a " cosmopolitan species" by the fact that it inhabits every continent of the world except Antarctica. As their name implies, barn owls are often found nesting in barn lofts, providing them both protection from the elements as well as a resident **prey base.** They may also choose to live in a variety of natural cavities, including cliff ledges, hollow trees and caves (Fig. 7).

Figure 7. Common barn owl nest and roost sites.

Appearance

The dorsal side of this beautiful bird of prey is colored golden-rust with numerous black flecks, while the face and entire **ventral** side varies from tan to white, lightening with age. The white facial disk is undoubtedly the most distinctive feature of the barn owl's appearance, strikingly heart-shaped framing black eyes and a sharp, downward-pointing beak (Fig.8). The barn owl is equipped with four clawed toes on each foot. When relaxed on a perch, three toes point forward and one backward. The front middle toe possesses a sharp pointed **talon,** the inner edge of which is uniquely supplied with fine serrations. It is believed that while the bird is **preening** this toe is used in a comb-like fashion to arrange and add loft to feathers of the facial disk. The outer toe on each foot is remarkably mobile, and in fact, is drawn to the rear a fraction of a second before the bird pounces on its prey. The barn owl's grasp seldom loses its prey, and has sufficient strength to kill mice almost instantly.

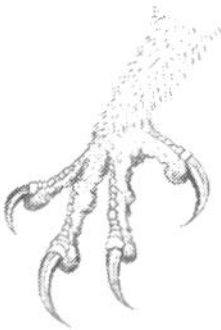

Figure 8. The barn owl facial disk and foot are specialized for its role as a nocturnal predator.

Barn Owl Description & Adaptations

Habits and Characteristics

The barn owl is a **nocturnal** animal, being most active at night and resting at a **roost** during the day. The facial disk serves to direct all available light toward the eyes of the barn owl. Contrary to popular myth, owls can see very well during the day as well as at night. Because the eyes are set in the front of the head both looking forward, owls have binocular vision, an adaptation crucial to accurate prey location. More importantly, the facial disk directs sound toward the ears of the barn owl. The barn owl can catch its prey in total darkness, using only its sense of hearing. Sound is so important to the barn owl that the flight feathers of this bird are modified with a velvety dorsal surface and a comb-like fringe on their leading edge to silence the flow of air as the owl glides through the night in search of a meal (Fig. 9).

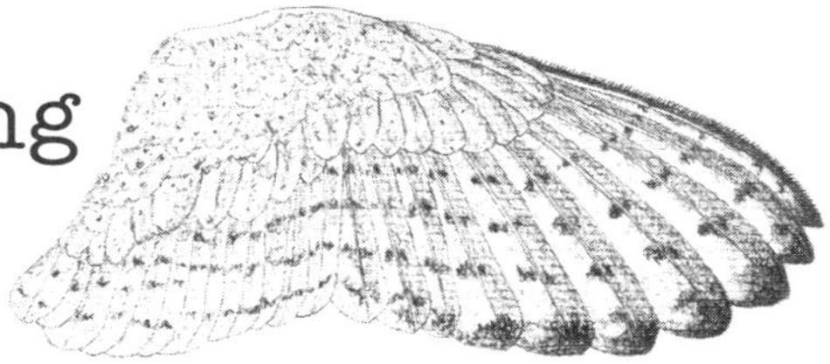

Figure 9.
The barn owl's wing has a velvety dorsal surface and a comb-like leading edge to reduce flight noise.

With the onset of the breeding season, barn owls become more active and participate in courtship behaviors which help to strengthen the pair bond. Barn owls mate for life and generally occupy a territory of approximately 2. 5 kilometers in the Pacific Northwest. Depending on the availability of food, the female may lay a large number of eggs, as many as eleven, whereas the average clutch size is closer to six eggs. After hatching, it takes about seventy-five days for a young owl to mature and leave the nest. The parents may nest again and produce a second **clutch** of eggs as long as the prey base will continue to support them. On the other hand, a poor "mouse year" may result in a depressed reproductive rate for the pair. The barn owl is also noted for its unusual practice of beginning incubation with the first egg laid. Most birds do not begin to incubate until the clutch is complete. Since each barn owl egg is laid about three days apart, if five eggs are laid, there may be two weeks age difference between the youngest and the oldest chicks.

Survival

As a predator dependent on successful hunting, life for the barn owl can be very difficult. About three quarters of all barn owls die in their first year. They may fly into wire or be hit by a car, they may drown, be poisoned or shot. However, most young barn owls die simply from starvation. Though not exceptionally large or fast in flight, barn owls have evolved to take advantage of the resources made available to them, whether in food or shelter, and have been known to eat anything from beetles to young rabbits as well as fish or small reptiles. Because this owl will make an attempt to catch almost anything in the appropriate size range, it may be labeled an **opportunist.** Hunting in almost total darkness and using its keen hearing to locate prey, the owl may catch almost any small mammal it encounters, a small weasel or squirrel for example. Owls that live near blackbird or starling roosts often eat large numbers of birds. A barn owl is not fast or agile enough to catch a small bird in flight, but is able to ambush these animals at their roost. In fact, barn owls do specialize in taking certain species of prey where and when that prey item is abundant enough to make the effort worthwhile.

Prey Species

As mentioned earlier, the apparent worldwide prey species of choice for **Tyto alba** is the predominant meadow vole of each respective region, commonly known as the field mouse (Fig.10). In fact, 90% of the barn owl's diet consists of this species. And, as you will see, nearly all of the barn owl's diet is derived from catching and eating small mamals.

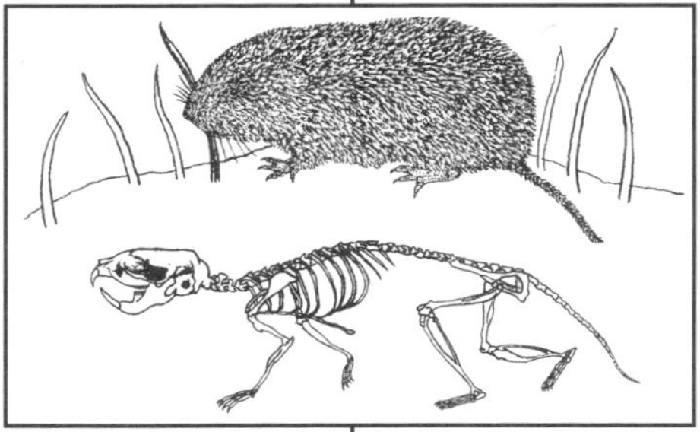

Figure 10. The meadow vole and its skeleton.

Microtus

Microtus is the genus name for all animals in this group and the species, or type, varies with the habitat and geographical range. Voles live in fields and along ditch rows, just about anywhere there is adequate food and abundant foliage to avoid predation. They are most actively grazing during the hours around sunrise and sunset.

An examination of the teeth of the vole will reveal information about its lifestyle. Possessing sharp chisel-like incisors, but lacking canines, the meadow vole is recognized as a **herbivore.** It chops down grass stalks in much the same way that a beaver fells trees, then uses its broad molars for grinding up the plant material into a digestible form (Fig. 10).

The meadow vole is also an opportunist with respect to reproduction. Unlike most of the small mammals it resembles, the vole is not locked into a seasonal breeding cycle. It may reproduce whenever conditions are favorable. This often produces wide fluctuations in the availability of prey for the barn owl. The barn owl in turn takes advantage of vole blooms by laying a greater number of its eggs and increasing its population size. When conditions change, however, so that the **habitat** can no longer support a large number of voles, the vole population crashes and the barn owl population usually does the same.

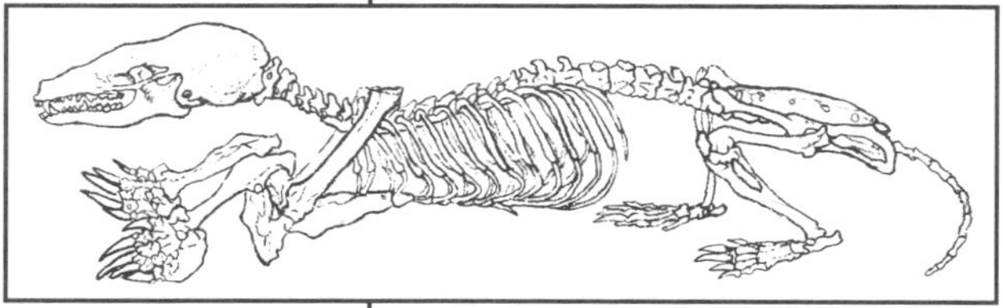

Figure 11.The mole skeleton

Prey Species

Scapanus

A closer look at the skeleton of the mole, **Scapanus** sp., will tell us more about another possible menu item in the barn owl's diet (Fig. 11). Mole bones will be found in about one out of fifteen pellets. The sturdy bones and claws of the forelegs tell us that they are used for heavy work burrowing underground in search of food (Fig. 12). Because of the mole's **fossorial** habits, it is less vulnerable to owl predation than other small mammals. Using their excellent sense of smell, moles hunt worms, various insects and their **larvae.** By examining the skull, one can see that the elongated nasal cavity provides a large surface area for olfactory receptors. The jagged teeth are used for capturing and chewing the small animals on which this **insectivore** preys.

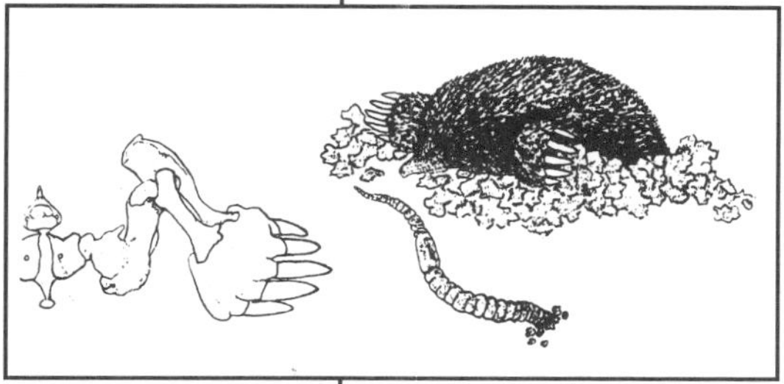

Figure 12.
The mole and it's specialized forelimb.

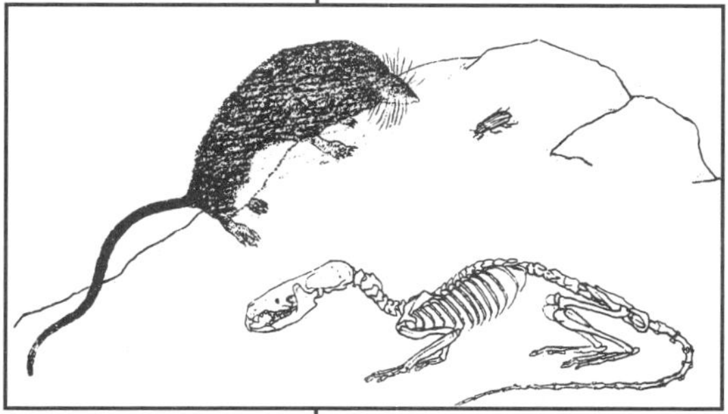

Figure 13.
The shrew and it's skeleton.

Sorex

Shrew skeletons are also commonly found in owl pellets, but because of their small size and delicate structure they are easily overlooked. The shrew, **Sorex** sp., is the smallest of all mammals, and a ferocious predator (Fig. 13). It has a fearsome set of teeth, and like the mole, an excellent sense of smell. In addition, the shrew secretes a poison with its saliva which serves to immobilize it's prey. Predators such as house cats often find shrews distasteful due to the presence of this chemical. However, the barn owl simply swallows shrews whole and shows no evidence of ill effects.

Figure 14.
A starling and a bird skeleton. Note large fused sternum and keel to anchor flight muscles, fused pelvic region and a beak rather than teeth for lightness.

Sturnis vulgaris & Passer domesticus

Bird skeletons which may be found in some pellets are very distinctive (Fig. 14). They reveal the special structural adaptations which enable flight. Their design has been molded by nature to provide support with a minimum of weight. The bones are hollow and many are fused together, to provide the strength and rigidity necessary for flight. The large breastbone serves as the anchor for the muscles that power the wings.

An examination of the beak can provide information about the diet and habitat in which the bird lives. For example, the starling, **Sturnis vulgaris,** has a long tapered beak for probing into soil for insects. (See Picture Key, Page 20). They spend much of their time foraging in open fields and lawns. The English sparrow, **Passer domesticus,** on the other hand, has a short but stout beak that is used for cracking seeds or grains.

Exoskeleton

In addition to mammals and birds, barn owls may also prey upon animals which possess an **exoskeleton.** The hard outer shell of these insects and crustaceans, made primarily of a substance called chitin, provides protection and multiple points of attachment for muscles. If a barn owl has eaten one of these armored animals, the meat will be digested and the exoskeleton will be regurgitated in the pellet.

Note: See the Picture Key, pages 18-20 for a more detailed explanation of the various skeletal dimensions of the many prey species of the barn owl.

Comparative Skeletal Anatomy

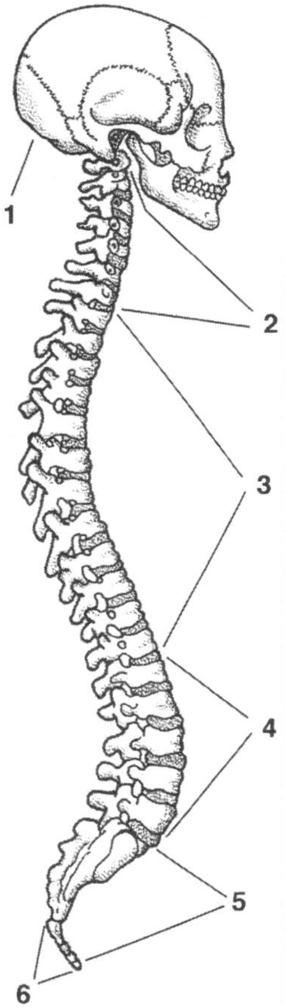

Figure 15. Human vetebral column.

1. Skull
2. Cervical
3. Thoracic
4. Lumbar
5. Sacrum
6. Coccyx

Animals such as barn owls, their prey, and humans have an endoskeleton of bone and cartilage surrounded by soft tissues. This skeleton is living tissue which grows and repairs itself, provides structural support and protection for vital organs, as well as enables movement by functioning as a sturdy system of jointed levers powered by contracting muscles. Animals such as insects and crayfish secrete a non-living material from their bodies which hardens into an exoskeleton. This provides for movement and the greatest possible protection, but restricts growth (animals with an exoskeleton must molt in order to grow), cannot be repaired and limits the range of motion. Imagine yourself encased in a suit of armor!

Most vertebrates, animals having an endoskeleton with a vertebral column, share the same basic skeletal pattern. The axial skeleton consists of the skull, rib cage, and vertebral column with five distinguishable regions: cervical, thoracic, lumbar, sacral and caudal (Fig. 15). The appendicular skeleton includes the two upper appendages (in our case, arms) each anchored to the axial skeleton by a pectoral girdle (scapula and clavicle), and two lower appendages (legs) each anchored to the axial skeleton by a pelvic girdle (pelvis).

Further, the number and position of bones is surprisingly consistent among mammals of different sizes and specializations for various lifestyles. The number of cervical (neck) vertebrae, for example, in the hamster is the same as in the sperm whale; and the lower forelimb of mammals always possesses a radius and an ulna, while the upper forelimb always possesses one bone, the humerus.

All animals have slightly different requirements of their skeleton depending on their particular way of life. In the mole for example, the humerus is comparatively short and stout with large crests for the attachment of digging muscles. The radius and ulna position the mole's palm outward from the body, and the bones of the front foot are large and flattened to resemble a shovel (Fig. 10). The front foot of a mole is more than twice as large as the back foot!

Comparative Skeletal Anatomy

The vertebral column may be likened to an archer's bow: strong but flexible and capable of transferring power as it changes shape. The lumber vertebrae of the meadow vole are long and narrow to increase flexibility and speed while running. By flexing its back, the vole adds push to its already pumping hindlegs during a sprint for cover. The mole's vertebral column is also highly flexible, but for a different reason. Its flexibility and shorter cervical region enable the mole to turn around easily in the cramped quarters of its burrow.

A brief comparison of the teeth and skull of various animals will reveal different adaptations for feeding and receiving sensory information from the environment. Rodents like the meadow vole have a pair of very distinctive upper and lower incisors. These long curved teeth continue to grow throughout life and have tooth enamel only on their outer surface to maintain a "self-sharpening" cutting edge. Look for the characteristic orange color of the enamel on the front of each incisor.

The presence of grinding molars and the lack of canines (creating a space called the diastema) indicate that the vole is a herbivore, and does not use its teeth in predatory behavior. In addition, the diastema allows for more side-to-side jaw movement during chewing. Carnivores, on the other hand, require a more stable jaw with upper and lower teeth which fit closely together for grasping and tearing meat, and the number of molars is reduced. Shrews and moles, both insectivores, have many small pointed teeth with very little difference between incisors, canines and premolars. As specialized carnivores, their dentition is related to a diet which requires them to break into the hard exoskeletons of insects.

The inside of the nasal cavity provides the surface area for olfactory (smell) receptors. Comparing the skulls found in an owl pellet, one can easily see by the large size of their nasal cavity that the sense of smell is very important to the mole and shrew.

Comparative Skeletal Anatomy

As a final comparison, the skulls of rodents possess a large zygomatic arch creating a deep channel through which the masseter (chewing) muscle passes. In contrast, the zygomatic arch is much reduced in the mole, and absent in the shrew.

To accomodate our large brain, humans have the largest cranial cavity for our body size, and comparatively smaller facial features. Our nose, eyes and mouth are positioned on the same vertical plane, giving roughly equal importance to the senses of smell, sight and taste. Our teeth give evidence as to our unspecialized omnivorous diet, and our dominant place in the food web. We exploit all levels of the food chain, eating a variety of both plants and animals. Our anatomy reflects the fact that, unlike smaller mammals, we rely less on our instinctive physical reactions to sensory information, and depend heavily on our ability to learn and manipulate the environment in order to survive.

Conservation

Efforts are now being made to promote the growth of barn owl populations where they have dwindled. The most successful method has been to provide suitable nest boxes, places well out of reach of potential predators such as cats, raccoons, opossums and weasels. A productive nest box provides ample opportunity to observe and study barn owls, as well as a predictable supply of pellets for dissection. Attention must be paid however, so that the birds are never disturbed at the nest site.

Almost any large wooden box equipped with an entrance hole of about seven inches in diameter will serve as a nest box for a pair of barn owls. For minimal investment in time and materials, the barn owl repays the biologist or casual observer with endless hours of fascination and enjoyment; and for the farmer a reward equal to the lion's share of captured mice and rats. A diagram is included in this booklet of a proven nest box design (Fig.16). Conservation efforts such as this acknowledge our appreciation for the diversity of life, as well as the wonder we have for each individual.

Important Facts to Consider

Barn owl nest boxes can be especially useful on farms and ranches where rodent control is desired.

About 85% of barn owl nesting attempts produce fledgling young. Reasons for mortality include human disturbance, limb breakage, and attacks by raccoons, opossums, skunks, cats, and bobcats.

Summer is the best time to erect a barn owl nest box. Boxes should not be disturbed during the nesting season or owls may desert them.

Honeybees frequently take over barn owl nest boxes, which make them useless for nesting. About 1/4 of the nest boxes in coastal southern California are used by honeybees.

Building a Nest Box

Barn owl nest boxes are easy to build. Ideal material is 3/8-inch or 1/2-inch plywood. Nest boxes should be painted using a camouflage pattern of drab green, black and brown to reduce human disturbance. Painting also helps prevent warping. Place a 2- inch layer of sawdust or wood chips in the bottom of the box and replace each year.

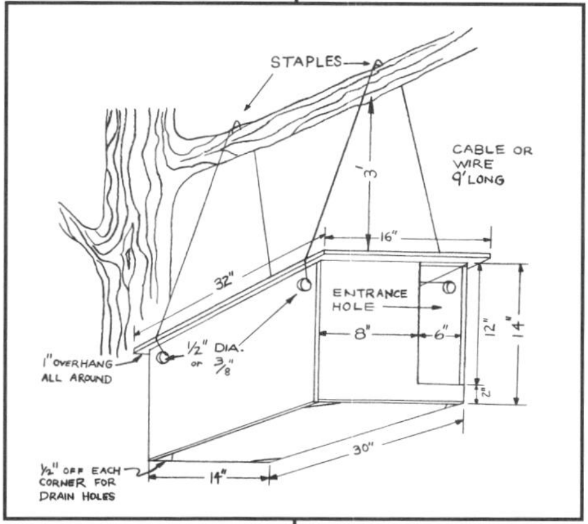

Guide

1. Make the entrance hole six inches wide by 12 inches high.
2. Hang or mount the box 10 to 30 feet above the ground and, if in a tree, 5 feet below a sturdy branch.
3. Use scrap exterior grade 3/8 inch to 1/2 inch plywood.
4. Use #4 or #5 galvanized hot dipped box nails or wood screws.
5. Use marine grade plastic resin or exterior wood glue for assembly.
6. Use wire staples to keep the cable from moving on the branch. Only necessary if the branch is not horizontal.
7. Use at least a 9 foot long piece of wire to hang boxes.

Figure 16.
A barn owl nest box.
(Courtesy of Pete Bloom, National Audubon Society).

Locating & Positioning a Nest Box

Good locations for barn owl nest boxes are wooded areas or open fields and meadows with a few trees. The box can face any direction. It can be three feet below a stout tree limb and be suspended by cables or mounted on poles 10 to 30 feet above ground. The entrance should be unobstructed and the box positioned so it does not attract human attention. Place about 6 boxes per square mile. Boxes mounted high in barns or abandoned buildings also are readily occupied by owls. If the nest box is placed on a post, the post should be wrapped with an 18" metal, conical predator guard.

Vole Skull.

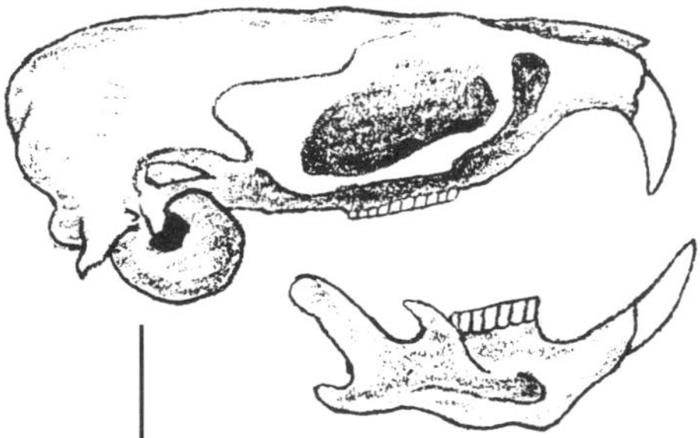

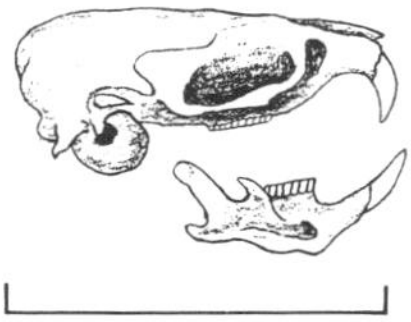

Actual Size 25 mm.

Average length 25-31 mm. Molars without roots. Molar configuration most important in vole identification. **Dental formula** 1-0-0-3/1-0-0-3 (upper jaw/lower jaw).

Rat Skull.

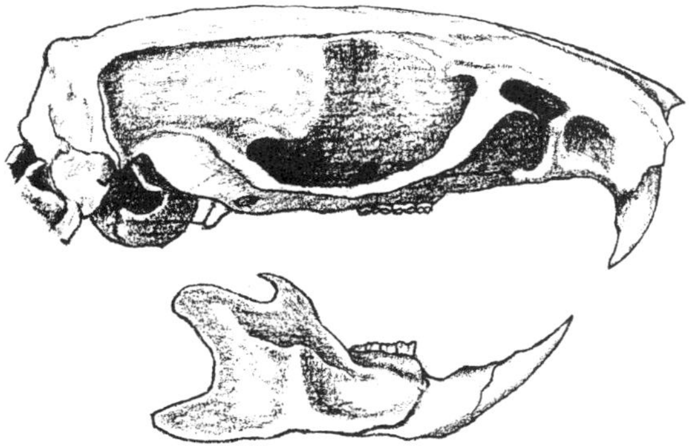

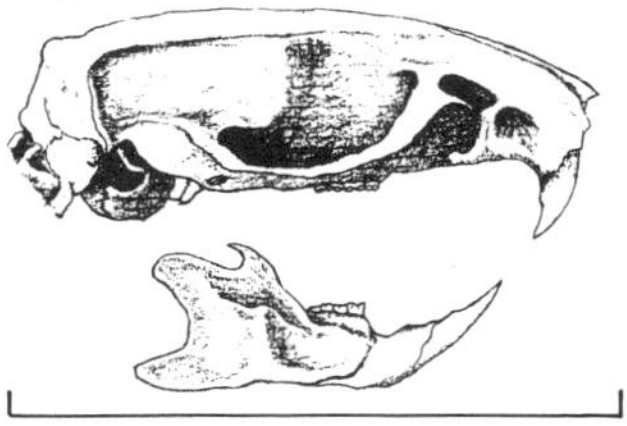

Actual Size 40 mm.

Average length 40-45 mm. Skull similar to vole with three rows of cusps on the molars. **Dental formula** 1-0-0-3/1-0-0-3 (upper jaw/lower jaw).

Picture Key

Weasel Skull.

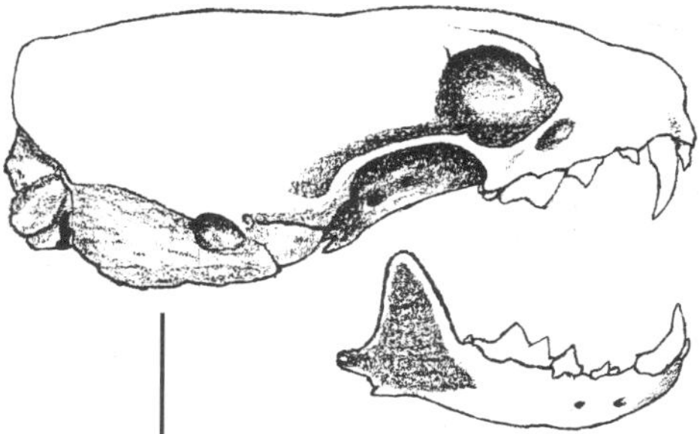

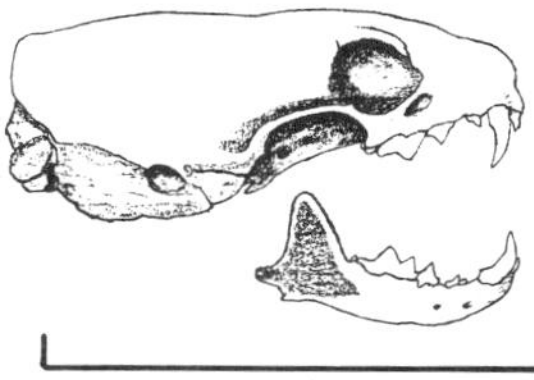

Actual Size 34 mm

Average length 29-44 mm. Species dependent. Note prominent canines. **Dental formula** 3-1-3-1/3-1-4-2.

Mole Skull.

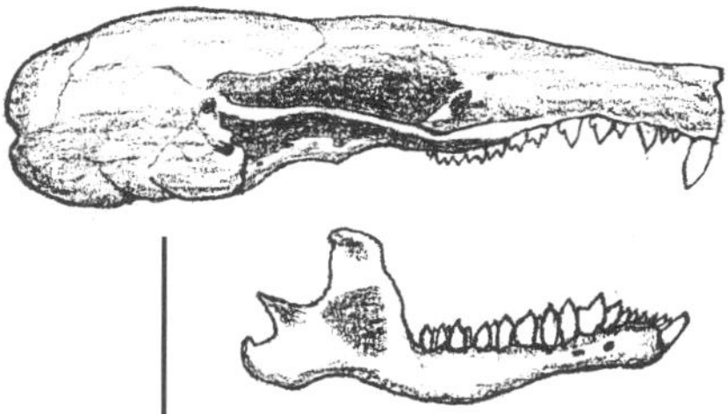

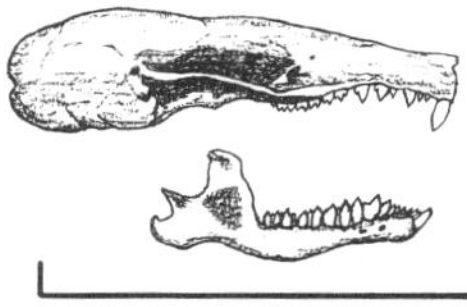

Actual Size 30 mm

Average length 29-44 mm. Species dependent. Weak zygomatic arch. **Dental formula** 3-1-4-3/ 3-1-4-3.

Shrew Skull.

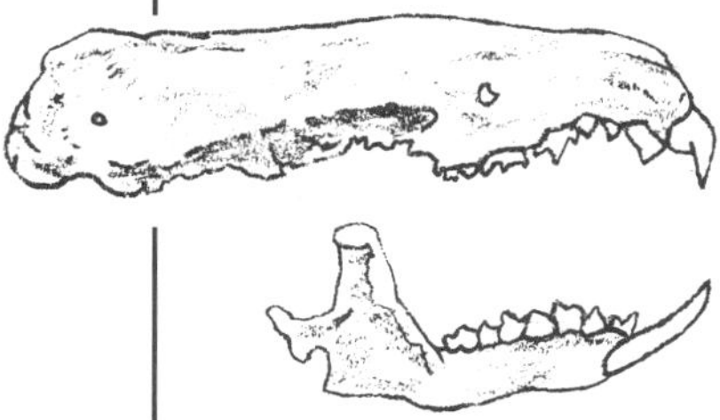

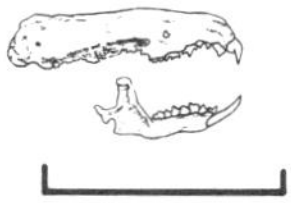

Actual Size 16 mm

Average length 16-22 mm. **Dental formula** varies between species; note red tipped teeth.

Picture Key

Starling Skull.

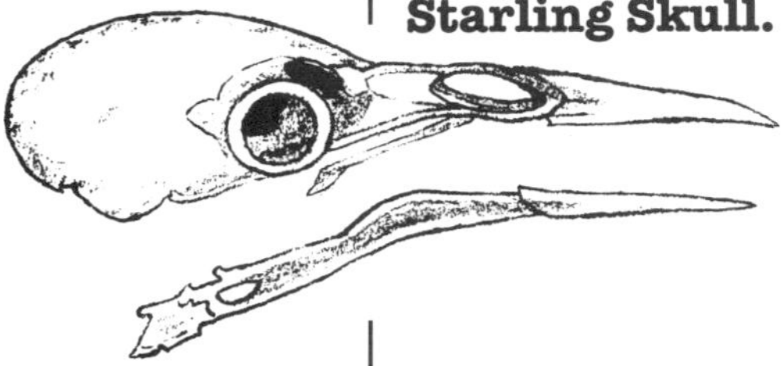

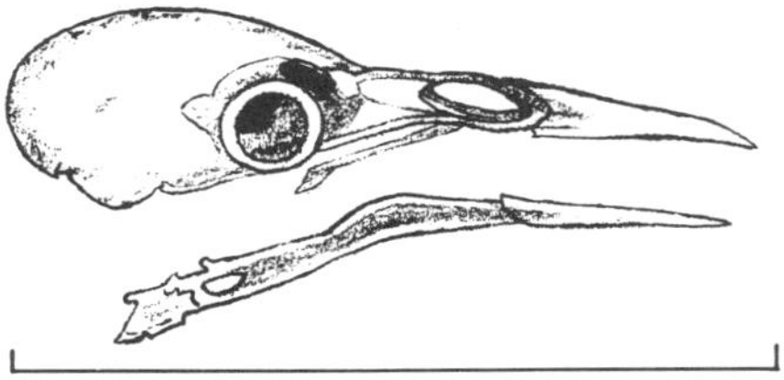

Actual Size 55 mm.

Average length 55 mm. Note long probing beak.

Blackbird Skull.

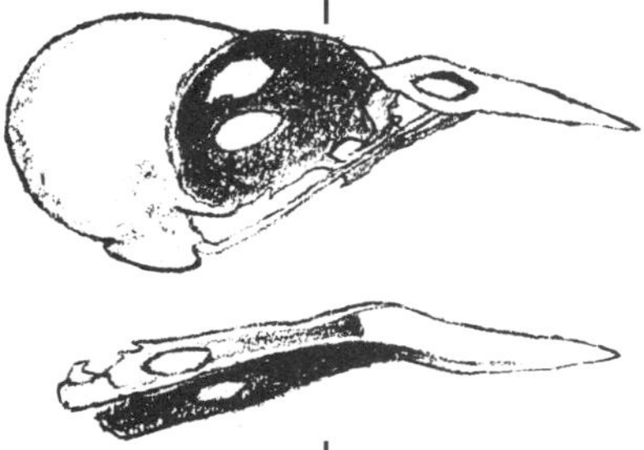

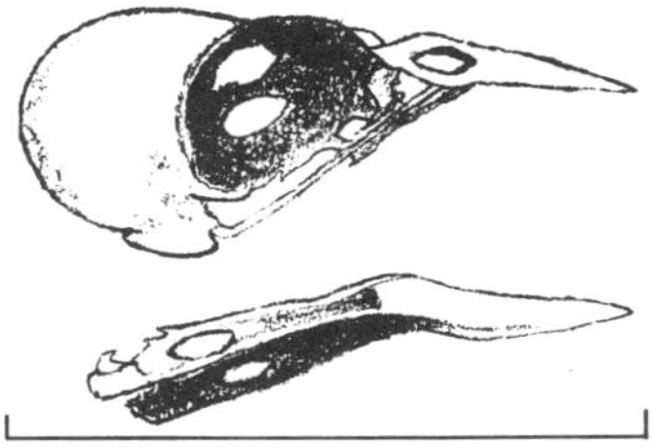

Actual Size 43 mm.

Average length 43 mm.

Sparrow Skull.

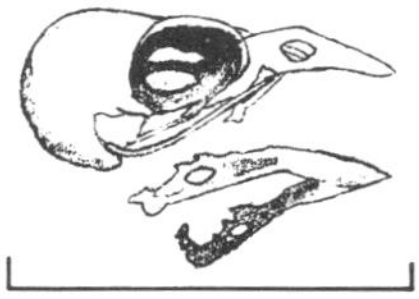

Actual Size 27 mm.

Average length 27-30 mm. Seed cracking beak.

Grasshopper Parts.

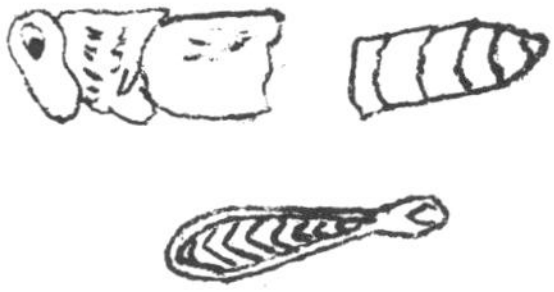

Review Questions

1. What is an owl pellet?

2. What constitutes the definition of a "predator" ?

3. What is a food chain?

4. Draw a diagram of your food chain.

5. How does solar energy make food?

6. What does a trophic pyramid represent?

7. What is a biological indicator? Is the barn owl? Are we?

8. Name some other nocturnal animals.

9. What is territory? Why are they important for predators?

10. How is the barn owl adapted to its role as a nocturnal predator?

11. What is a herbivore? Can you tell from the teeth?

12. How many differences between a bird and a mammal can you name?

13. Is a barn owl good or bad? Why?

Glossary

Anterior: Toward the front, or head region of an organism; the opposite of posterior.

Autotroph: An organism capable of extracting energy and raw (inorganic) materials from the environment and manufacturing its own food (organic molecules).

Biological indicator: A species of animal or plant which because of their role in the ecosystem may be used to evaluate the health of that community or organisms and their environment.

Biomass: The total weight or calorie (energy) value of a living organism.

Bios: Life, living.

Biosphere: The thin ring of atmosphere, water and soil, at the surface of the earth in which all organisms live.

Carnivore: An animal which satisfies its energy needs by feeding on animal flesh.

Casting: A matted ball of undigested fur, feathers and bone regurgitated by all birds of prey.

Chitin: A strong, flexible chemical substance (polysaccharide); the principle component of the exoskeleton of insects and crustaceans.

Clutch: The eggs and chicks produced in one breeding effort.

Dental Formula: Represents one side of the jaw. First number = incisors, second number = the number of canines, third number = pre-molars and the fourth number = number of molars.

Dorsal: Toward the back; in chordates, on the same side as the spinal column; the opposite of ventral.

Ecology: The study of the interactions of organisms with their controlling environment and with each other, and the influence such interactions have on the ecosystem.

Glossary

Energy: In biological systems, the power to do work when given appropriate form (i.e. chemical energy, solar energy); may be lost from an ecosystem in the form of heat.

Exoskeleton: An external skeleton, providing support, protection of internal organs, and enabling movement.

Food Chain: An orderly sequence of organisms which indicates the direction of the flow of energy and raw materials in the ecosystem.

Food Web: The complex arrangement of interlocking food chains in a community.

Foraging: The search for food.

Fossorial: Living underground.

Genus: A category in the binomial system of classification which groups organisms closely related to each other; usually includes several different species.

Gizzard: The muscular organ of a bird's digestive system which prepares food for entry into the stomach; the site of pellet formation.

Habitat: The characteristic place (geographic location and climate) in which an organism lives.

Herbivore: An organism which meets its energy requirements by feeding on plant tissue.

Incubation: Maintaining eggs at body temperature for normal development and hatching.

Insectivore: An organism which feeds on insects.

Larva: A newly hatched immature form of animal capable of feeding; especially in insects.

Mammal: A classification of warm-blooded animals which are covered with hair or fur, and which possess milk-producing (mammary) glands.

Glossary

Metabolism: All of the chemical reactions within a cell or organism; usually controlled by a series of enzymes.

Nocturnal: Active at night.

Olfactory: Sense of smell.

Opportunist: An organism which adjusts its actions to take advantage of circumstances.

Posterior: Toward the back or tail region of an organism; the opposite of anterior.

Predator: An animal which kills and eats other animals.

Preen: In birds, using the beak to clean and arrange the feathers.

Prey: The animal hunted and eaten by particular predators.

Roost: A place which provides a safe perch for birds to rest.

Solar: Of the sun.

Species: A distinct group of similar organisms whose genetic make-up and behavior allow them to interbreed and produce fertile offspring.

Talon: The claw on each toe of a bird of prey.

Trophic Ecology: The study of the flow of energy between groups of organisms in an ecosystem.

Trophic Pyramid: A diagram representing the transfer of energy from autotrophs at the base through succeeding levels of herbivores, carnivores, and decomposers.

Ventral: Toward the front; in chordates, on the underside, opposite of dorsal.

Selected Readings

Booth, E.S. **How to Know the Mammals** (1971, William C. Brown Pblg.) The Picture Key Nature Series. An easy to use picture key to the mammals, including glossary. 198 pp.

Bunn, D.S., A.B. Warburton, R.D.S. Wilson **The Barn Owl** (1982, Buteo Books Vermillion South Dakota). Forty years of fieldwork with excellent photos and interesting facts. The most recent and the best reference. 264 pp.

Burton J.A. ed. **Owls of the World** (1973, EP. Dutton). Their evolution structure and ecology. A beautiful picture book encyclopedic in nature. 216 pp.

Clark, R.J. **Working Bibliography of Owls the World** (1978, Scientific and Technical series 1.)

Earle, O.L. **Mice at Home and A field** (1957, William Morrow & Co., N.Y) A charming and informative story of the life of a mouse; illustrated. 63 pp.

Glass, B.P **A Key to the Skulls of North American Mammals** (1961, Burgess Publishing Co.) A well illustrated but fairly technical working key. 4 pp.

Ingless, L.G. **Mammals of the Pacific States** (1986, Stanford University Press) Illustrated with over five hundred photographs drawings and maps. 606 pp.

Kritzman, E.B. **Little Mammals of the Pacific Northwest** (1977, Pacific Search Press). A nice little book with species accounts and natural history, well illustrated with some photos. 120 pp.

Melanby, K. **The Mole** (1973, Tepltnger Publishing) Short and good reading life history.

Oxford Scientific **Harvest Mouse** (1982, G.P. Putnam's Sons N.Y.) Films. A beautiful little picture book on the life of this mouse. 20 pp. Elementary.

Ripper, C.L **Moles and Shrews** (1967, Morrow Junior Books, William Morrow and Company). A delightful look at the lives of these two small mammals, very fully illustrated. 84 pp. Elementary.

Silverstein, A. & V.B. **Rats and Mice, Friends and Foes of Man** (1988) Fact and folklore, illustrated with many photos. 92 pp. Elementary.

Selected Readings

Sparks, K.
Soper, T.
Owls: Their Natural and Unatural History (1970, Taplinger Publishing). Entertaining combination of folklore and fact; well illustrated. 206 pp.

Tyler, T.A.
Philips, D.
Owls by Day and By Night (1978, Naturegraph Publications). Excellent introduction to the North American Owls, both general and specific content; full of illustrations and photos. 208 pp.

Whitaker, J.O.
The Audubon Society Field Guide to the North American Mammals (1980, Borzoi Book, Alfred A. Knopf, Inc.). Standard field guide information with color photos of animals in natural settings. 746 pp.

List of Figures

Note: 8x10 transparencies are available for all figures. (except #15)

Notes

Notes